Name: Freya

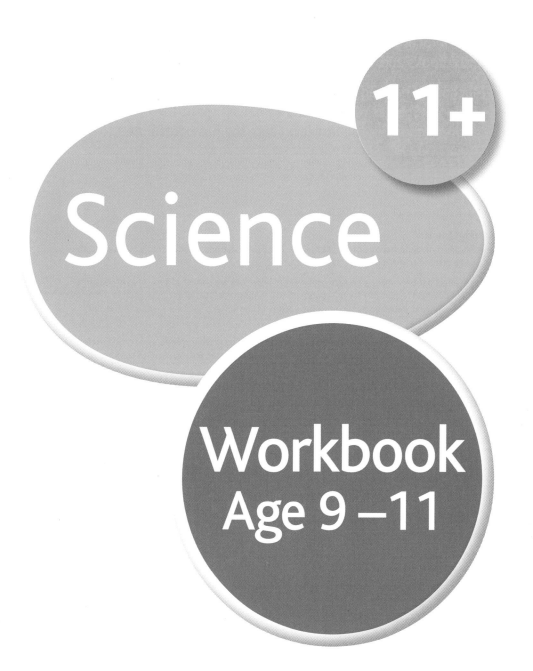

11+
Science
Workbook
Age 9–11

Sue Hunter
Jenny Macdonald

About the authors

Sue Hunter has very recently retired as a full-time science teacher but she continues to be very involved in Science education. She is a member of the Common Entrance setting team and a governor of local primary schools. Sue has written extensively, including Galore Park's KS2 Science textbooks and 11+ Science Revision Guide.

Jenny Macdonald has had a happy and fulfilling career as a teacher, teaching in both state and private schools. She has recently retired having spent the last eighteen years teaching science in a local prep school. In the last few years she has contributed to Galore Park's KS2 Science textbooks.

The Publishers would like to thank the following for permission to reproduce copyright material.

Photo credits

p28 (top) © Everett Historical / Shutterstock (bottom) © Chronicle / Alamy Stock Photo

Acknowledgements

Every effort has been made to trace all copyright holders, but if any have been inadvertently overlooked, the Publishers will be pleased to make the necessary arrangements at the first opportunity.

Although every effort has been made to ensure that website addresses are correct at time of going to press, Galore Park cannot be held responsible for the content of any website mentioned in this book. It is sometimes possible to find a relocated web page by typing in the address of the home page for a website in the URL window of your browser.

Hachette UK's policy is to use papers that are natural, renewable and recyclable products and made from wood grown in sustainable forests. The logging and manufacturing processes are expected to conform to the environmental regulations of the country of origin.

Orders: **Teachers** please contact Bookpoint Ltd, 130 Park Drive, Milton Park, Abingdon, Oxon OX14 4SE. Telephone: (44) 01235 400555. Email: primary@bookpoint.co.uk. Lines are open from 9 a.m. to 5 p.m., Monday to Saturday, with a 24-hour message answering service.

Parents, Tutors please call: 020 3122 6405 (Monday to Friday, 9:30 a.m. to 4.30 p.m.). Email: parentenquiries@galorepark.co.uk

Visit our website at www.galorepark.co.uk for details of other revision guides for Common Entrance, examination papers and Galore Park publications.

ISBN: 978 1 5104 2981 9

© Sue Hunter and Jenny Macdonald 2018

First published in 2018 by
Hodder and Stoughton Limited,
An Hachette UK Company
Carmelite House
50 Victoria Embankment
London EC4Y 0DZ

www.galorepark.co.uk

Impression number   10 9 8 7 6 5 4 3 2

Year   2022 2021 2020

All rights reserved. Apart from any use permitted under UK copyright law, no part of this publication may be reproduced or transmitted in any form or by any means, electronic or mechanical, including photocopying and recording, or held within any information storage and retrieval system, without permission in writing from the publisher or under licence from the Copyright Licensing Agency Limited. Further details of such licences (for reprographic reproduction) may be obtained from the Copyright Licensing Agency Limited, www.cla.co.uk

Illustration on p33 by Oxford Designers & Illustrators.

Illustrations on the following pages by Aptara, Inc.: p7, p9 (top), p10, p19, pA6, pA7 (all), p51 (top), p53 (top), p59, p60 (all), p62 (all)

Illustrations on all other pages by Integra Software Services Ltd.

Typeset in India

Printed in India

A catalogue record for this title is available from the British Library.

Contents and progress record

Use these pages to record your progress. Tick the 'Completed' box when you are confident with each skill and record your scores in the 'Score' boxes.

- Introduction 6

1 Classification and keys

Topic	Page	Completed	Score
What I already know	7	☐	/ 12
Sorting living things: an owl is like a tiger	8	☐	/ 10
Sorting living things: insect and spider	9	☐	/ 14
Identification keys: fungi	10	☐	/ 5
Identification keys: plant identification	11	☐	/ 4
Classification and keys: crossword	12	☐	/ 11
Quick quiz: classification and keys	13	☐	/ 18

2 Healthy living

Topic	Page	Completed	Score
What I already know	15	☐	/ 10
Nutrients: what's on my plate?	16	☐	/ 11
Nutrients: testing for starch	17	☐	/ 10
Transporting substances around the body: heart rate and exercise	18	☐	/ 10
Taking a deep breath	19	☐	/ 15
Activity for health	20	☐	/ 9
Drugs	21	☐	/ 7
Quick quiz: healthy living	22	☐	/ 25

3 Microbes

Topic	Page	Completed	Score
What I already know	24	☐	/ 6
Microbes and disease	25	☐	/ 10
Helpful microbes: yeast investigation	26	☐	/ 14
Research topic: the history of science	28	☐	/ 10
Quick quiz: microbes	29	☐	/ 18

4 Evolution and inheritance

- What I already know .... 31 — / 9
- The history of life on Earth .... 32 — / 4
- Evolution .... 33 — / 10
- Making new varieties of plants and animals .... 34 — / 16
- Evolution: puzzle .... 35 — / 11
- Quick quiz: evolution and inheritance ..... 36 — / 13

5 Hot and cold

- What I already know .... 37 — / 7
- Measuring temperature .... 38 — / 10
- Keeping warm: potatoes .... 39 — / 14
- Keeping warm: house insulation .... 40 — / 12
- Research topic: polar bears .... 41 — / 10
- Quick quiz: hot and cold .... 42 — / 12

6 Indicators

- What I already know .... 44 — / 9
- Indicators: wordsearch .... 45 — / 10
- Testing common substances .... 46 — / 11
- Making an indicator .... 47 — / 10
- Quick quiz: indicators .... 48 — / 14

7 Light and sight

- What I already know .... 50 — / 10
- Light and shadows .... 51 — / 17
- Reflection .... 53 — / 12
- Sight .... 54 — / 6
- Light and sight: crossword .... 55 — / 11
- Quick quiz: light and sight .... 56 — / 16

8 Changing electrical circuits

	Completed	Score
What I already know 58	☐	☐ / 15
Changing circuits: graphic organiser 59	☐	☐ / 10
Changing circuits: odd one out.................................. 60	☐	☐ / 6
Drawing circuit diagrams 61	☐	☐ / 15
Short circuits... 62	☐	☐ / 10
Changing electrical circuits: puzzle 63	☐	☐ / 12
Quick quiz: changing electrical circuits 64	☐	☐ / 11

- Answers   A1

Introduction

This workbook is divided into eight chapters matching Galore Park's Science Year 6 textbook, so you can use it alongside the textbook or on its own.

Each chapter begins with a 'What I already know' section, to remind you about the topic and to provide you with a quick check of your prior learning before you start work. You will then have the opportunity to practise a wide range of scientific skills, including:

- identifying and describing similarities and differences
- labelling a diagram neatly
- using a number key to identify organisms
- identifying questions to make a branching key
- developing your scientific vocabulary
- understanding data displayed in graphs and pie charts
- communicating clearly in words and pictures
- plotting graphs and bar charts

and so on.

You can easily identify the skill you are going to practise by reading the 'Skill' box at the top of each page.

The 'Skill check' box at the end of each page will encourage you to assess your confidence with each skill.

Each chapter finishes with a quick quiz and a check list to help you assess your progress through the 11+ syllabus and National Curriculum content.

The **contents and progress record** (pages 3–5) helps you keep track of your progress. When you are confident you have mastered each skill, tick the 'Completed' box and record your total score in the 'Score' box.

You might find yourself using this book in the classroom, or for homework, or you might work through it independently at home. It is linked broadly to the 11+ Common Entrance examination syllabus and the National Curriculum, but it will help you to develop key skills and reinforce your learning if you are preparing for any of the following tests:

- Common Entrance 11+ in Science
- 11+ entry tests set by individual schools
- end-of-year tests
- pre-tests.

Answers are provided in a pull-out section in the middle of the book.

1 Classification and keys

What I already know

1 Using what you have already learnt, draw neatly ruled lines to complete each sentence.

To help us study living things we sort them into	mammals.
Animals with internal skeletons are called	flowers.
Slugs, snails, worms, insects and spiders are	invertebrates.
Plants are living things that contain	amphibians.
Plants are sorted according to whether they have	vertebrates.
Fungi are not plants because they do not carry out	chlorophyll.
Animals with fur on their bodies are	groups.
Newts belong to the group called	photosynthesis.

(8)

2 Which groups do these animals belong to?

Frog _____

Grasshopper _____

Snake _____

Parrot _____

(4)

Sorting living things: an owl is like a tiger

Skill Identifying and describing similarities and differences

Look carefully at the pictures and identify features to help you to compare each pair of animals. Think what else you know about these animals. Describe the similarities and differences clearly.

1 An owl is like a tiger because:

   (i) _____

   (ii) _____

   (iii) _____

   (iv) _____

   (v) _____ (5)

2 An owl is different from a robin because:

   (i) _____

   (ii) _____

   (iii) _____

   (iv) _____

   (v) _____ (5)

Skill check

I have identified key features of the animals. ☐

I have described the similarities and differences clearly. ☐

Sorting living things: insect and spider

Skill Labelling a diagram neatly

Use words from the box below to label the pictures of the spider and the insect. Remember to arrange your labels neatly at the sides of the pictures and draw labelling lines carefully using a pencil and ruler.

abdomen	head	thorax	3-part body
antennae	insect	wings	3 pairs of legs
eyes	spider	2-part body	4 pairs of legs

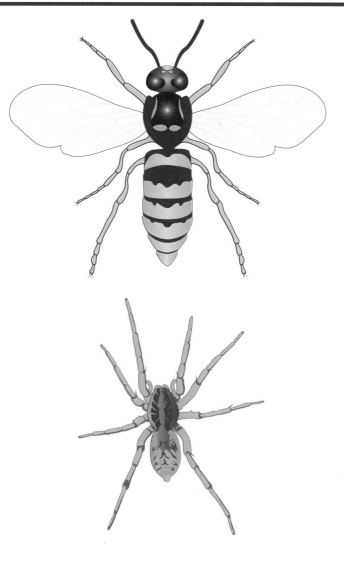

(14)

Skill check

I have written the labels neatly.

I have used a pencil and ruler to draw accurate labelling lines.

Identification keys: fungi

Skill Using a number key to identify organisms

Use the number key to identify the fungi in the pictures and fill the correct names into the table at the bottom of the page.

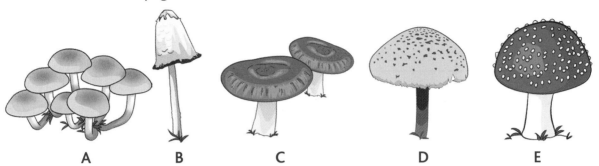

A   B   C   D   E

1   The cap has spots on it   Go to 2
    The cap has no spots on it   Go to 3

2   The cap has white spots on it   FLY AGARIC
    The cap has brown spots on it   PARASOL

3   The fungi grow in a cluster   SULPHUR TUFT
    The fungi do not grow in a cluster   Go to 4

4   The fungi have red caps   THE SICKENER
    The fungi have white caps   SHAGGY INK CAP

	Name of fungus
A	
B	
C	
D	
E	

(5)

Skills check

I observed diagnostic features of the fungi. ☐

I know how to use a number key to identify organisms. ☐

Identification keys: plant identification

Skill Identifying questions to make a branching key

The key below does not have any questions in it. Write the questions below in the correct boxes to complete the key. One question is missing. You will need to make up your own question to complete the key to identify the plants in the pictures.

Are the flowers brightly coloured?
Does the plant have flowers?
Is the leaf made up from small leaflets?

Grass   Dog Rose

Seaweed

Buttercup

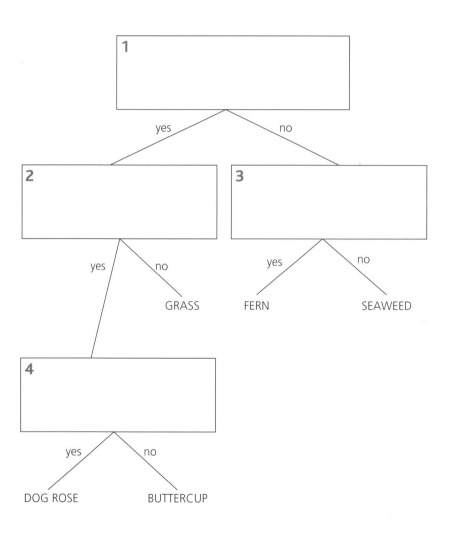

Fern

(4)

Skills check

I understand how to use questions to make a branching key. ☐

I can write questions that have YES or NO as the answer. ☐

Classification and keys: crossword

Skill Developing vocabulary

Across

6 An owl is a bird because it has these on its body
7 Animals with internal skeletons
8 An animal that has fur on its body and feeds its young on milk
9 The group of animals that includes frogs, toads and newts
10 An invertebrate with 4 pairs of legs
11 A trout is a fish because it breathes through these

Down

1 The type of feature that can be used to distinguish between organisms for classification
2 Ferns, mosses and algae belong to the group of non-_____ plants
3 The group of animals with dry, scaly skin that lay soft-shelled eggs on land
4 The group containing all invertebrates with jointed legs
5 Fungi are not classified as plants because they do not contain this pigment

(11)

Skill check

I know and understand the important vocabulary in this crossword.

Quick quiz: classification and keys

Write *short* answers to the following questions to check how much you remember about classification and keys. You do not need to write full sentences.

1 Give one way in which the work of Carl Linnaeus helps scientists around the world to study living things.

_____ (1)

2 What diagnostic feature distinguishes plants from animals?

_____ (1)

3 What is the key difference between vertebrates and invertebrates?

_____ (1)

4 Give the name of the group of invertebrates that have jointed legs.

_____ (1)

5 What are the two main differences between insects and spiders?

_____
_____ (2)

6 How many vertebrate groups are there?

_____ (1)

7 How do mammals feed their young?

_____ (1)

8 Give one way in which you can tell that a bat is not a bird.

_____ (1)

9 Give one key difference between reptiles and amphibians.

_____ (1)

10 Give two features of fish that are not normally found in other animals.

_____ (2)

11 Give an example of a non-flowering plant.

_____ (1)

12 Name three groups of organisms that are not classified as animals or plants.

_____ (3)

13 Why are fungi not classified as plants?

_____ (1)

14 Why is size not a diagnostic feature?

_____ (1)

(Total 18)

Check list: classification and keys

I know	😊	😐	☹
why it is important to sort plants and animals into groups			
about the work of Carl Linnaeus			
that plants and animals are classified into separate kingdoms			
how to distinguish between vertebrates and invertebrates			
that arthropods are invertebrates with jointed legs			
the key differences between insects and spiders			
that vertebrates are divided into five groups			
the distinguishing features of mammals, birds, fish, amphibians and reptiles			
that plants are classified as either flowering or non-flowering			
that bacteria, fungi and single-celled organisms are not part of the plant or animal kingdoms			
why fungi are not included in the plant kingdom			

I can	😊	😐	☹
identify which features of an organism are diagnostic and which are not			
design and use simple branching keys			
design and use simple number keys			

2 Healthy living

What I already know

The table below contains words that you should know from your earlier work about how your body works.

Write words or phrases in the central box in each row to link the words to make scientifically accurate sentences.

Blood		heart.
Carbohydrates		energy.
Fats		energy.
Oxygen		lungs.
Proteins		growth.
Food		stomach.
Delicate organs		skeleton.
Sugars		decay.
Vitamin C		scurvy.
A balanced diet		nutrients.

(10)

Nutrients: what's on my plate?

Skill Understanding data displayed in a pie chart

The picture shows a plate of food.

The table shows some nutrients that are contained in the food on this plate.

	Mass of nutrient, in g				
Food	Protein	Fat	Fibre	Carbohydrate	Salt
Fish finger	7.0	4.0	1.0	10.0	0.3
Chips	2.2	7.5	4.0	34.0	0.4
Peas	4.6	0.5	3.5	7.0	0.0
Ketchup	0.2	0.0	0.0	3.5	0.3
Total					

1 Calculate the total mass of each nutrient contained in this plate of food. Write the totals into the table. (5)

2 The pie chart below shows the proportions of the nutrients in the plate of food. Use your calculations to help you to label the pie chart to show which nutrient is represented by which segment. (5)

Pie chart to show the proportion of the nutrients in the plate of food

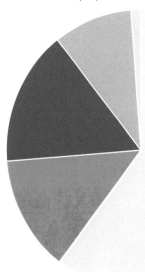

3 The data above do not include all the nutrients contained in the plate of food. Give one group of nutrients that is not represented. _____ (1)

Skill check

I understand how to interpret data displayed in a pie chart. ☐

Nutrients: testing for starch

Skill Communicating clearly in words and pictures

Some children are given several foods. Their teacher asks them to find out which of the foods contain starch.

Use the spaces and lines below to create a clear set of illustrated instructions showing the steps that they should take to test all the foods for starch.

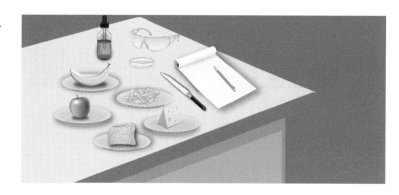

(10)

Skill check

I have explained the steps in the starch test clearly.
My illustrations are neat and help my explanation.

Transporting substances around the body: heart rate and exercise

Skill Plotting a graph

Gloria is a keen rower. Faye, who does not row very often, challenges her to a rowing race. They wear sensors to measure their pulse rates before, during and after the race.

The table shows their pulse rate results.

Rower	Pulse rate, in beats per minute				
	Resting rate	At end of race	1 minute after the race	5 minutes after the race	10 minutes after the race
Gloria	68	140	110	75	68
Faye	72	165	140	105	90

1. Use the data from the table to show Gloria's pulse rates at the end of the race and 1, 5 and 10 minutes after the race on the graph below. Do not include her resting rate. Draw a smooth curve to join the points. Label this line with Gloria's name. (5)

2. Use the data from the table to show Faye's results in the same way on the graph. Draw a smooth curve to join the points. Label this line with Faye's name. (5)

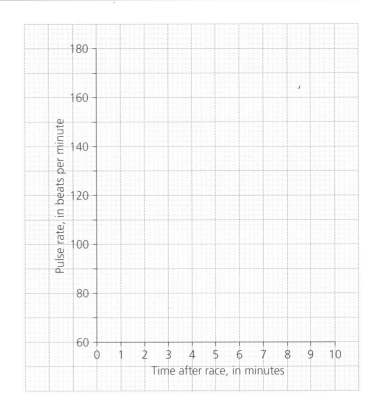

Skill check

I have plotted the points on the graph neatly and accurately. ☐

I have drawn neat, smooth curves to join the points. ☐

Taking a deep breath

Skill Developing vocabulary and interpreting data

1 Complete the following passage using words from the box below.

oxygen	vein
heart	glucose
surface area	breathing
artery	respiration
lungs	carbon dioxide

The muscle that pumps the blood is called the _____. The blood is pumped away from the heart through an _____ to the _____ and returns to the heart through a large _____. It is then pumped strongly around the body, carrying _____, _____ and water to the body cells.

Air is taken in and out of the lungs. This is called _____. Inside the lungs, oxygen passes into the blood and _____ is removed from the blood. The surface of the lungs is greatly folded to give a large _____ for gas exchange.

The body cells take in oxygen and glucose and use these to provide energy for activity. This is called _____. (10)

2 When we take exercise, our pulse rate and breathing rate change.

By referring to information from the table or the graph on the previous page, describe three differences in the pulse rates of the two rowers.

_____
_____
_____
_____ (3)

3 Suggest an explanation for the differences you have described.

_____
_____ (2)

Skill check

I know and understand the words used in this exercise. ☐

I can apply my knowledge to explain differences in data. ☐

Activity for health

Skill Applying knowledge

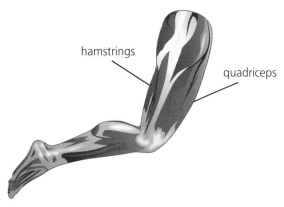

■ Muscles in the leg

1 The diagram shows the leg of a footballer who is about to kick a ball. Use your knowledge of how the elbow works to help you to suggest how the labelled muscles worked together when she bent her knee ready to kick the ball.

   _____

   _____ (2)

2 Explain how the muscles will work to straighten her leg when she kicks the ball.

   _____

   _____ (2)

3 What name is given to pairs of muscles working together in this way?

   _____ (1)

4 Using the words *obesity*, *stamina*, *strength* and *exercise*, write a sentence explaining how playing a sport, such as football, regularly, helps people to be healthy.

   _____

   _____ (4)

Skill check

I have used my knowledge of the muscles in the elbow to work out how a knee joint works. ☐

Drugs

Skill Evaluating information in a text and applying knowledge

Read the two short passages below.

A: In 1908, in a book called *Scouting for Boys: A Handbook for Instruction in Good Citizenship*, Lord Baden-Powell gave the following advice:

'A scout does not smoke… He knows that when a lad smokes before he is fully grown up it is almost sure to make his heart feeble… Any scout knows that smoking spoils his eyesight, and also his sense of smell, which is of greatest importance to him for scouting on active service.'

B: In October 1914, at Lord Kitchener's request, a Smokes for Soldiers and Sailors Fund was to be formed 'to provide our wounded … with tobacco and cigarettes in hospitals here and at the front … and is at the moment sending regular supplies to over 200 hospitals and convalescent homes.' (Quote from *The Times* newspaper, 29 October 1914)

1 Which of these two passages best fits with modern advice on smoking? Explain your answer.

_____

_____

_____

_____

_____

_____ (5)

2 Suggest reasons why a young soldier in the trenches in 1914 might choose to smoke, even though he had read Lord Baden-Powell's advice in 1908.

_____

_____

_____ (2)

Skill check

I have read the information in the two passages carefully. ☐

I have used my knowledge to evaluate the information in the passages. ☐

I have explained my ideas clearly, with reference to the texts. ☐

Quick quiz: healthy living

Write *short* answers to the following questions to check how much you remember about healthy living. You do not need to write full sentences.

1 Give two nutrient groups that provide the body with energy.
   _____ (2)

2 Which chemical is used to demonstrate the presence of starch in a food?
   _____ (1)

3 Why are proteins needed in the diet?
   _____ (2)

4 What is the function of the heart?
   _____ (1)

5 Name two types of blood vessel.
   _____ (2)

6 Name two substances transported around the body by the blood.
   _____ (2)

7 Which gas is absorbed into the blood from the air?
   _____ (1)

8 How are the lungs adapted to help this gas to enter the bloodstream?
   _____ (1)

9 How might the lungs of a smoker differ from those of a non-smoker?
   _____ (1)

10 Why does pulse rate increase during exercise?
   _____ (2)

11 Give three health benefits of regular exercise.
   _____ (3)

12 Give three functions of the skeleton.
   _____ (3)

13 What is meant by the term 'antagonistic muscles'?
   _____ (2)

14 What is meant by the term 'addictive substance'?
   _____ (1)

15 Give one reason why people must not drink alcohol before driving a vehicle.
   _____ (1)

(Total 25)

Check list: healthy living

I know	🙂	😐	🙁
that food provides nutrients for energy, growth and repair			
that carbohydrates are energy-providing foods and include starch and glucose			
that proteins are needed for growth and repair			
that fats store energy and are needed for insulating the body			
that the heart is a muscle that pumps blood around the body, including the lungs			
how the blood flows around the body through arteries and veins			
that oxygen, nutrients, water and carbon dioxide are transported around the body in the blood			
that during breathing, oxygen is taken into the lungs and carbon dioxide is removed from the lungs			
that the lungs have a large surface area through which gases can be exchanged between air in the lungs and the blood			
that smoking reduces the surface area of the lungs, leading to severe breathing difficulties			
why the body needs more oxygen and food (glucose) during exercise			
why pulse rate increases during exercise			
that exercise builds stamina and muscle strength and helps to prevent obesity			
that respiration is the release of energy from food (glucose)			
how the skeleton provides support and protection and aids movement			
how joints are moved by pairs of antagonistic muscles working together			
that smoking may cause lung diseases such as cancer, and heart disease			
that tobacco contains nicotine, which is an addictive drug			
that alcohol slows reaction times and may damage the liver			
that solvents and other illegal drugs affect body organs, especially the brain			

I can	🙂	😐	🙁
carry out the iodine test for starch			
draw a pie chart from given information			
draw and interpret a line graph			

3 Microbes

What I already know

The boxes below each contain half a sentence. Draw lines to join the pairs to make full sentences.

| We should always wash our hands | | so other people do not catch our colds. |

| | Doctors give people inoculations (injections) | |

| before eating, after using the toilet and after handling animals. | | to stop them catching certain diseases. |

| | We put used tissues in the bin | Some illnesses can be cured |

| Kitchen surfaces should be kept clean | break down and recycle waste in the environment. | |

| Decomposers, such as fungi and bacteria, are living things that | because raw foods may contain things that could make us ill. | by medicines called antibiotics. |

(6)

Microbes and disease

Skill Developing scientific vocabulary

Hidden in the wordsearch are ten words linked to the topic of microbes and disease.

There is a clue to each word below the word gird. Highlight each word in the wordsearch and write it on the answer line by each clue.

Q	L	C	O	Y	E	A	S	T	D	A	H
T	X	H	H	Y	G	I	E	N	E	N	D
K	P	Y	V	S	A	M	P	D	C	T	M
A	F	F	I	S	J	Z	C	D	O	I	I
Y	H	I	R	S	E	T	O	J	M	B	C
N	C	F	U	W	N	Y	M	W	P	I	R
G	S	C	S	D	N	P	O	L	O	O	O
Q	M	J	U	M	E	H	U	A	S	T	B
U	N	X	R	N	R	O	L	O	E	I	E
H	F	J	C	Y	N	I	D	E	R	C	M
R	B	C	P	N	H	D	S	S	S	R	Q
G	L	F	O	B	A	C	T	E	R	I	A

Clues

1 Another name for a micro-organism _____

2 Food poisoning is caused by microbes from this group _____

3 The doctor who discovered a way of preventing smallpox _____

4 A medicine that may be used to cure diseases caused by bacteria _____

5 The process of keeping ourselves and our environment clean _____

6 Flu is caused by this type of micro-organism _____

7 A type of fungus used in baking and brewing _____

8 Micro-organisms that break down waste in the environment _____

9 A type of fungus that may grow on foods if not stored properly _____

10 A bacterial disease that can be spread through infected drinking water

_____

(10)

Skill check

I know and can use the vocabulary from this topic. ☐

Helpful microbes: yeast investigation

Skill Using investigative techniques and interpreting data

Some pupils want to investigate whether temperature affects how quickly bubbles are produced by yeast.

They use the following apparatus.

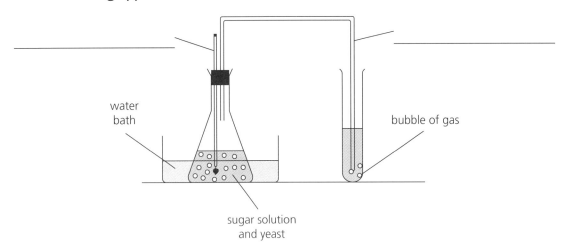

water bath

bubble of gas

sugar solution and yeast

The pupils place yeast and sugar solution in the flask. They can control the temperature of the experiment using the water bath.

They count how many bubbles are released by the yeast mixture in one minute.

1 Complete the labelling on the diagram. (2)

2 Suggest two things they should do to make their investigation a fair test.

  _____

  _____ (2)

3 Suggest what they should do to make sure that these results are as reliable as possible.

  _____

  _____ (2)

4 Suggest why it is important to have a closely-fitting stopper in the top of the flask.

  _____

  _____ (2)

Their results are shown on the graph below.

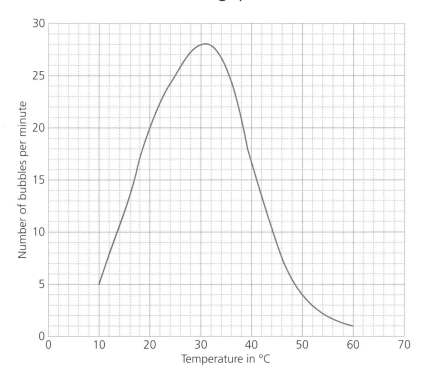

5 Describe what these results tell us about how temperature affects the rate at which bubbles are formed by yeast.

   _____

   _____ (2)

6 Use the graph to estimate how many bubbles the children counted at 40 °C. Show your working on the graph.

   _____

   _____ (2)

7 How many bubbles would you expect to be produced at 70 °C? Explain your answer.

   _____

   _____ (2)

Skills check

I can explain a pattern in a set of data shown on a graph. ☐

I can describe how to carry out a fair test and ensure that results are reliable. ☐

Research topic: the history of science

Skill Using books and the internet for research

Keeping a bibliography

There are lots of people who have helped us to understand microbes and disease. Two people whose work has played a large part in helping to prevent disease are Louis Pasteur and Joseph Lister.

Louis Pasteur

Louis Pasteur showed that microbes in the air are responsible for making wine turn sour. The process he invented to keep wine fresh is still used today to keep many products, including milk, free of disease. Describe this process.

■ Louis Pasteur

_____

_____

_____

_____

(5)

Joseph Lister

Joseph Lister was a surgeon who was worried by how often people died from infections after surgery. He read a work by Pasteur that suggested that microbes in the air could cause disease and infection. What did he do to keep his patients safe?

■ Joseph Lister

_____

_____

_____

_____

(5)

Bibliography

List all the books and websites you used to answer the questions about Pasteur and Lister.

_____

_____

_____

_____

Skill check

I have found reliable sources to research the work of these two scientists. ☐

I have recorded my sources of information in a bibliography. ☐

Quick quiz: microbes

Write *short* answers to the following questions to check how much you remember about microbes. You do not need to write full sentences.

1. What is a micro-organism (microbe)?

   _____ (1)

2. Name three types of micro-organism.

   _____ (3)

3. What is meant by the term 'decomposer'?

   _____ (1)

4. Which micro-organism is important in making bread, wine and beer?

   _____ (1)

5. Name a disease caused by a fungus.

   _____ (1)

6. Name two diseases caused by bacteria.

   _____ (2)

7. Name two diseases caused by viruses.

   _____ (2)

8. Explain why your doctor will not give you antibiotics if you have a cold.

   _____
   _____ (2)

9. What is the name of the process, invented by Jenner, that is used to prevent diseases such as smallpox?

   _____ (1)

10. Why should you always wash your hands before eating a meal?

    _____
    _____ (2)

11. Why is it important to keep fresh foods, such as meat and dairy products, in the fridge?

    _____
    _____ (2)

(Total: 18)

Check list: microbes

I know	😊	😐	😞
that a micro-organism (microbe) is a living thing that is too small to see without a microscope			
that the three main types of micro-organism are bacteria, viruses and fungi			
that micro-organisms can be helpful or harmful			
that decomposers are micro-organisms that break down waste and recycle nutrients in the environment			
that yeast is a micro-organism that is used in making bread, wine and beer			
that diseases caused by fungi include athlete's foot and ringworm			
that diseases caused by bacteria include sore throats, whooping cough, bronchitis, food poisoning and typhoid			
that diseases caused by viruses include colds, flu, measles and chicken pox			
that antibiotics are medicines that that help the body's natural defences (immune system) to fight diseases caused by bacteria			
that diseases caused by viruses cannot be treated with antibiotics			
that vaccination is a method of disease prevention that protects against diseases such as smallpox, polio, and typhoid			
that it is important to wash hands before meals, after using the toilet and after handling animals to prevent the spread of diseases			
that hygiene in hospitals and food preparation areas is essential to prevent the spread of disease			
that micro-organisms grow more quickly in warm, moist conditions			

I can	😊	😐	😞
interpret information shown in a graph			
describe the steps needed to carry out a fair test			
describe how to ensure that results from an investigation are reliable			

4 Evolution and inheritance

What I already know

Choose words or phrases from the box to complete the following sentences using your knowledge of rocks and fossils.

dead	lakes or seas	reptiles
different	millions of years	sedimentary
hard	plant	skeletons and teeth

1 Fossils are the remains of _____ animals and plants. (1)

2 Fossils are found in _____ rocks. (1)

3 Fossils are usually formed from the _____ parts of animals. (1)

4 Fossils are formed when the remains of dead organisms fall to the bottom of _____ and become covered in sediments. (1)

5 Fossils are the remains of organisms that lived _____ ago. (1)

6 Fossilised organisms are often very _____ to those alive today. (1)

7 Dinosaurs were large _____ that lived millions of years ago. (1)

8 Coal is a solid fossil fuel made from _____ remains. (1)

9 By looking at _____, scientists can predict information about the lifestyle of a fossil animal. (1)

The history of life on Earth

Skill Interpreting information in text

Read the following passage carefully.

Over millions of years, the horse has changed as its habitat changed. This process is known as evolution, and we can trace it through the fossil record. Two ancestors of the horse were *Hyracotherium* and *Merychippus*. The diagrams show some ways in which the modern horse evolved.

	Hyracotherium	*Merychippus*	*Equus* (modern horse)
What the animal may have looked like	height: 0.4 m	height: 1 m	height: 1.5 m
Bones in the foot			

60 million years ago, the land was marshy and dotted with bushy vegetation. *Hyracotherium* was about as big as a small dog. Its feet had four widely spread toes that allowed it to move quickly across the marshy ground to avoid predators.

Over time, the marshy areas became drier and covered in open grassland, with fewer bushy places to hide. Running on firmer ground is easier on fewer toes. By 20 million years ago the horse, *Merychippus*, had just three toes as the two central toes had become fused together. This process continued and by about 3–4 million years ago, horses had developed a single hoof on each foot.

Use information from the passage above to answer the following questions.

1. Suggest why a foot with widely-spread toes might have been beneficial to *Hyracotherium*.
   _____ (2)

2. Over time, the feet of the horse ancestors changed. What other important change in the animals is shown in the diagrams above?
   _____ (1)

3. *Hyracotherium* has been drawn with a spotted coat. Explain why we cannot be certain what the coat of these animals looked like.
   _____ (1)

Skills check

I read and understood the information in the Text.

I used my knowledge and understanding to suggest explanations.

Answers

1 Classification and keys

What I already know (page 7)

1 To help us study living things we sort them into **groups**.
   Animals with internal skeletons are called **vertebrates**.
   Slugs, snails, worms, insects and spiders are **invertebrates**.
   Plants are living things that contain **chlorophyll**.
   Plants are sorted according to whether they have **flowers**.
   Fungi are not plants because they do not carry out **photosynthesis**.
   Animals with fur on their bodies are called **mammals**.
   Newts belong to the group called **amphibians**. (8)
2 Frog – amphibians
   Grasshopper – insects
   Snake – reptiles
   Parrot – birds (4)

Sorting living things: an owl is like a tiger (page 8)

1 Possible similarities include:
   They are both vertebrates.
   They are both carnivores/predators.
   They are both warm-blooded.
   They both have forward facing eyes.
   They both have sharp teeth/claws. (5)
   (Allow valid alternatives – 1 mark for each.)
2 Possible differences include:
   The owl has bigger eyes than the robin.
   The owl's eyes are at the front of the head but the robin's are on the side.
   The owl is nocturnal but the robin is active during the day.
   The owl is bigger than the robin.
   The owl hunts small mammals but the robin hunts insects. (5)
   (Allow valid alternatives – 1 mark for each.)

Sorting living things: insect and spider (page 9)

- Labels on insect: insect, 3-part body, 3 pairs of legs, wings, head, thorax, abdomen, eyes, antennae
- Labels on spider: spider, 2-part body, 4 pairs of legs, eyes, head (14)

Identification keys: fungi (page 10)

A Sulphur tuft
B Shaggy ink cap
C The sickener
D Parasol
E Fly agaric (5)

Identification keys: plant identification (page 11)

1 Does the plant have flowers?
2 Are the flowers brightly coloured?
3 Is the leaf made up from small leaflets?
4 Pupil's own question to distinguish clearly between buttercup and dog rose. (4)

Classification and keys crossword (page 12)

Across

6 feathers
7 vertebrates
8 mammal
9 amphibians
10 spider
11 gills

Down

1 diagnostic
2 flowering
3 reptiles
4 arthropods
5 chlorophyll (11)

Quick quiz: classification and keys (page 13)

1 Use of Latin names in all languages avoids confusion when referring to species. (1)
2 Plants contain chlorophyll/carry out photosynthesis. (1)
3 Vertebrates have an internal bony skeleton/invertebrates have no bony skeleton. (1)
4 Arthropods (1)
5 Number of legs (insects 6, spiders 8), number of body parts (insects 3, spiders 2) (2)
6 Five (1)
7 Suckle with milk (produced by glands in the mother's body) (1)
8 Bats do not have feathers/birds do not have fur. (1)
9 Reptiles lay eggs on land, amphibians in water/reptiles have dry scaly skin, amphibians have moist skin. (1)
10 Gills and fins (2)
11 Fern/alga/seaweed/moss (1)
12 Fungi, bacteria (monera) and single-celled organisms (protists) (3)
13 They do not contain chlorophyll/carry out photosynthesis. (1)
14 Because animals and plants grow/differ in size due to environmental conditions. (1)
(Total 18)

2 Healthy living

What I already know (page 15)

Pupils should be able to make suitable connections, for example:

- Blood is pumped around the body by the heart.
- Carbohydrates are a useful source of energy.
- Fats are used by the body to store energy.
- Oxygen is taken into the body through the lungs.
- Proteins are needed by the body for growth.
- Food is digested in the stomach.
- Delicate organs are protected by the skeleton.
- Sugars in the diet can cause tooth decay.
- Vitamin C is needed to prevent scurvy.

- A balanced diet contains a balance of different nutrients. (10)
  (Accept valid alternatives.)

Nutrients: what's on my plate (page 16)

1. protein 14 g, fat 12 g, fibre 8.5 g, carbohydrate 54.5 g, salt 1 g (5)
2. Yellow = carbohydrate
   Pale blue = salt
   Dark blue = protein
   Orange = fat
   Grey = fibre (5)
3. Vitamins (1)

Nutrients: testing for starch (page 17)

The instructions should show the following stages:

- Cut/take a small piece of a food and place it on a clean petri dish or similar.
- Add drops of iodine solution.
- Observe results. If the food contains starch, the iodine will change from brown to blue–black.
- Record results in a table.

Additional information that could also be included:

- Wear eye protection.
- Clean apparatus between tests.
- Wash hands afterwards.

Illustrations should show each step clearly and support the written instructions. (10)

(Award 5 marks for correct method and 5 marks for illustrations.)

Transporting substances around the body: heart rate and exercise (page 18)

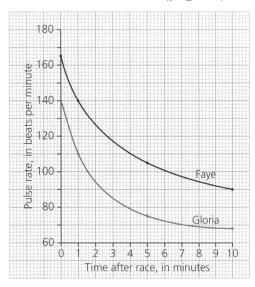

1. Gloria's four data points correctly plotted on the graph and neatly joined with a smooth curve and clearly labelled. (5)
   (4 for correctly plotted points. 1 for line)
2. Faye's four data points correctly plotted on the graph and neatly joined with a smooth curve and clearly labelled. (5)
   (4 for correctly plotted points. 1 for line)

Note: If the pupil is not familiar with drawing curves, the points could be joined by straight lines using a ruler.

Taking a deep breath (page 19)

1. The muscle that pumps the blood is called the **heart**. The blood is pumped away from the heart through an **artery** to the **lungs** and returns to the heart through a large **vein**. It is then pumped strongly around the body, carrying **oxygen**, **glucose** and water to the body cells. Air is taken in and out of the lungs. This is called **breathing**. Inside the lungs, oxygen passes into the blood and **carbon dioxide** is removed from the blood. The surface of the lungs is greatly folded to give a large **surface area** for gas exchange.
   The body cells take in oxygen and glucose and use these to provide energy for activity. This is called **respiration**. (10)
2. Gloria's resting pulse rate (68 bpm) is lower than Faye's (72 bpm).
   Faye's pulse rate rises further during exercise (increase of 93 bpm to 165 bpm compared with an increase of 72 bpm to 140 bpm).
   Gloria's pulse rate drops by 30 bpm in the first minute, but Faye's only drops by 25 bpm/the steeper line on the graph shows that Gloria's pulse rate drops more quickly than Faye's in the first minute.
   After 10 minutes Gloria's pulse rate has returned to her resting rate but Faye's is still above her resting rate. (Any three required) (3)
   (Note: In the question 'By referring to information from the graph or table…' indicates that values should be included in the answer.)
3. Gloria rows on a regular basis so her heart is less affected by the strenuous exercise/she is fitter than Faye. (2)

Activity for health (page 20)

1. To bend the knee, the quadriceps relax and the hamstrings contract to pull the lower leg backwards. (2)
2. To straighten the leg to kick the ball, the hamstrings relax and the quadriceps contract to pull the lower leg forwards. (2)
3. Pairs of muscles working together to move a joint are called antagonistic muscles. (1)
4. Regular exercise increases muscle strength and stamina, and helps to prevent obesity. (4)
   (1 mark for each word correctly used.)

Drugs (page 21)

1. Baden-Powell's advice is nearer to the current advice because he is telling scouts not to smoke, and he is correct that smoking affects the heart and sense of smell, although he does not include warnings about the damage to the lungs. (Extension: However, Baden-Powell is suggesting that it is only a bad thing for 'a lad to smoke before he is fully grown up', and modern advice is that smoking is harmful at any age.) (5)
   (Up to 5 marks, awarded for valid observations. Additional 'bonus' marks could be awarded at the discretion of the marker.)
2. Trench **warfare was stressful** and uncomfortable and soldiers often found that the **nicotine in cigarettes was calming and relaxing**. (2)
   (Allow any sensible suggestion clearly argued.)

Quick quiz: healthy living (page 22)

1. Carbohydrates and fats (2)
2. Iodine (solution) (1)
3. Growth and repair (2)
4. To pump blood around the body. (1)
5. Arteries and veins (allow capillaries) (2)
6. Oxygen, glucose/nutrients from digested food, carbon dioxide, water (2)
   (Any two required. Allow valid alternatives)
7. Oxygen (1)
8. They have a large surface area. (1)
9. Smokers' lungs have a reduced surface area/coated with tar, thus reducing effective surface area. (1)
10. To transport blood carrying oxygen and glucose more quickly to the muscles. (2)
11. Increased stamina, increased muscle strength, prevention of obesity (3)
12. **Protection** of delicate organs, **support** of the body, **attachment of muscles** for movement (3)
13. A pair of muscles **pulling in opposite directions** to **move a joint**. (2)
14. A substance that causes people to become dependent so that they **cannot give it up**. (1)
15. Alcohol slows reaction times/causes drowsiness/impairs judgement. (1)
    (Any one required)

(Total 25)

3 Microbes

What I already know (page 24)

The completed sentences should read:

- We should always wash our hands before eating, after using the toilet and after handling animals.
- Doctors give people inoculations (injections) to stop them catching certain diseases.
- We put used tissues in the bin so other people do not catch our colds.
- Some illnesses can be cured by medicines called antibiotics.
- Kitchen surfaces should be kept clean because raw foods may contain things that could make us ill.
- Decomposers, such as fungi and bacteria, are living things that break down and recycle waste in the environment. (6)

Microbes and disease (page 25)

1. MICROBE
2. BACTERIA
3. JENNER
4. ANTIBIOTIC
5. HYGIENE
6. VIRUS
7. YEAST
8. DECOMPOSERS
9. MOULD
10. TYPHOID (10)

Helpful microbes: yeast investigation (page 26)

1. **Thermometer** in the flask and **delivery tube** between the flask and the test tube correctly labelled. (2)
2. Use the same mass of yeast/same volume of water/same mass of sugar in each test. (2)
   (Any two required)
3. **Repeat the readings** (three times) and **calculate the mean** number of bubbles for each temperature. (2)
4. The stopper ensures that **all the gas produced** passes through the delivery tube to the test tube to **form bubbles that can be counted**. (2)
5. The number of bubbles produced per minute rises as the temperature rises until about 30 °C. Above this temperature, the rate drops again. (2)
6. 17 bubbles per minute. (1)
   Lines correctly drawn onto the graph to show working. (1)
7. At 70 °C it is likely that no bubbles would be produced because the yeast cells would be killed at this temperature. (2)
   (Give credit for alternative answers if clearly explained.)

Research topic: the history of science (page 28)

<u>Louis Pasteur</u>

Pasteur discovered that heating wine briefly to 55 °C killed bacteria without ruining the taste. This process was later named pasteurisation. To pasteurise milk, it is heated to 71.7 °C. The milk is kept at this temperature for at least 15 seconds and no longer than 25 seconds. It is then rapidly cooled. This kills harmful bacteria and so milk is safer to drink and lasts for longer without turning sour. (5)

<u>Joseph Lister</u>

After reading Pasteur's work he decided to try using a technique proposed by Pasteur. He soaked dressings in an antiseptic solution called carbolic acid (phenol) to kill the micro-organisms causing infection. This resulted in much lower infection rates in his patients. He also experimented with hand-washing, sterilising instruments and spraying the operating theatre with carbolic acid before surgery. (5)

Quick quiz: microbes (page 29)

1. A living thing that is too small to be seen with the naked eye. (1)
2. Bacteria, viruses, fungi (3)
3. An organism that breaks down and recycles waste material in the environment. (1)
4. Yeast (1)
5. Athletes' foot/ringworm (1)
   (Any one required. Allow valid alternatives.)
6. Sore throat/whooping cough/bronchitis/food poisoning/typhoid (2)
   (Any two required. Allow valid alternatives.)
7. Colds/flu/measles/chicken pox (2)
   (Any two required. Allow valid alternatives.)
8. Colds are caused by viruses and antibiotics are not effective at treating diseases caused by viruses. (2)
9. Vaccination (1)
10. To remove disease-causing microbes from the hands

so they do not get into the mouth. (2)
11 Microbes such as fungi/moulds and bacteria cause these foods to spoil and they may build up to levels that would cause food poisoning if not kept cool. (2)

(Total 18)

4 Evolution and inheritance

What I already know (page 31)

1 Fossils are the remains of **dead** animals and plants. (1)
2 Fossils are found in **sedimentary** rocks. (1)
3 Fossils are usually formed from the **hard** parts of animals. (1)
4 Fossils are formed when the remains of dead organisms fall to the bottom of **lakes or seas** and become covered in sediments. (1)
5 Fossils are the remains of organisms that lived **millions of years** ago. (1)
6 Fossilised organisms are often very **different** to those alive today. (1)
7 Dinosaurs were large **reptiles** that lived millions of years ago. (1)
8 Coal is a solid fossil fuel made from **plant** remains. (1)
9 By looking at **skeletons and teeth**, scientists can predict information about the lifestyle of a fossil animal. (1)

The history of life on Earth (page 32)

1 They would **spread the weight** of the animal to **stop it from sinking** into the marshy ground. (2)
2 The height increases over time. (Allow changes in coat patterns.) (1)
3 Fossils very rarely preserve evidence of soft tissue/skin type/colour, etc. (1)

Evolution (page 33)

The resulting picture should show that key processes have been considered and appropriate solutions imagined (e.g. what it eats and how it would catch/obtain the foods, how it would move around the various parts of the island, how and where it might breed and how big it might become). Learners might also have considered how the animal is adapted to cope with temperature differences and whether or not there are predators on the island, and if so how the animals would respond. Possible solutions might include camouflage, nocturnal behaviour, fur, webbed feet, etc. Key features should be clearly labelled.

Making new varieties of plants and animals (page 34)

1 (6)

	Hair shedding	Affectionate	Intelligence/ trainability	Barking	Sensitivity	Aggressive behaviour
Guide dog	Very low	Very	High	Quiet	Easy-going	Low

2 No, because the offspring vary due to inheriting different mixes of characteristics from the parents. (2)
3 (a) Good sense of smell/easy to train/keen to please/not aggressive (2)
(b) Affectionate/easy-going/trainable/good with children (2)
(c) Strong/good stamina/able to work in a team (2)
(d) Good sense of hearing/trainable/affectionate/docile/keen to please (2)
(Accept valid suggestions.)

Evolution: puzzle (page 35)

1 Anning
2 Species
3 Dinosaurs
4 Fossil
5 Evolution
6 Palaeontologist
7 Darwin
8 Asteroid
9 Extinct (9)
The word is geologist. A scientist who studies rocks. (2)

Quick quiz: evolution and inheritance (page 36)

1 **Fossil** evidence shows that living things alive **millions of years ago were very different** to those alive today. (2)
2 Fossils (1)
3 Mary Anning (1)
4 Charles Darwin and Alfred Russel Wallace (2)
5 Individuals with **longer necks will survive better** (because they are better able to reach the leaves). They will breed and **pass the characteristic on to their offspring**. Over many generations, **more tortoises will have longer necks**. (3)
6 Evolution (1)
7 They will each inherit different characteristics from each parent. (1)
8 Breeding plants or animals from **chosen parents** to achieve **particular characteristics**. (2)
(Total 13)

5 Hot and cold

What I already know (page 37)

- Boiling kettle: 100 °C
- Body temperature: 37 °C
- Ice lolly: 0 °C
- Hot oven: 200 °C
- Room temperature: 20 °C
- Freezer: −15 °C
- Fridge: 5 °C (7)

Measuring temperature (page 38)

1. A = 95 °C
   B = 190 °F
   C = 67 °C
   D = 104 °F
   E = 34 °C
   F = −14 °C
   G = −14 °F (7)
2. (a) Boiling point of water = 212 °F (1)
   (b) Freezing point = 32 °F (1)
3. −13 °C (1)

Keeping warm: potatoes (page 39)

1. Nothing: 43 °C
   Woolly scarf: 26 °C
   Bubble wrap: 25 °C
   Newspaper: 32 °C
   Hay: 37 °C (5)
   (1 for each correct answer.)
2. To act as a control/to compare with the other results. (1)
3. Same mass/size of potato (1)
4. Five correctly drawn bars on bar chart (allow error carried forward from Q1) (5)
5. Bubble wrap (1)
   The potato in bubble wrap lost the least amount of heat energy/temperature dropped least. (1)

Keeping warm: house insulation (page 40)

1. Sheep wool (1)
2. Straw (1)
3. Absorbs heat in warm weather and releases it when conditions cool. It will keep the house cool by day and warm the house by releasing heat overnight. (2)
4. Sheep wool, because it says that is long lasting. Cellulose fibre may settle, reducing air pockets and straw will decay. (2)
5. Underlined:
   - Cellulose fibre: recycled newspaper and card
   - Sheep wool: renewable/long lasting
   - Straw: renewable/provides habitat (allow widely available, as this reduces transport)
   Cross by:
   - Cellulose fibre: printing inks may release vapours/fire risk if not treated with fire-retardant chemicals
   - Sheep wool: sheep produce methane, which contributes to climate change
   - Straw: fire risk if not treated with fire-retardant chemicals (6)

Research project: polar bears (page 41)

Facts that might be included:
- layer of fat under skin
- black skin to absorb heat
- small ears to reduce heat loss
- thick, multi-layered fur traps air for insulation

Other relevant facts may be included.

To encourage good practice, all sources should be listed in the bibliography and findings written in their own words.

Quick quiz: hot and cold (page 42)

1. Thermal insulator (1)
2. They **trap air,** which is a **good insulator.** (2)
3. It stops heat energy entering to warm the inside. (1)
4. How hot or cold something is (1)
5. Thermometer (1)
6. °C (1)
7. 100 °C (1)
8. 0 °C (1)
9. 37 °C (1)
10. Loft insulation, double glazing, cavity wall insulation, lagging hot water pipes, curtains. (Any two valid answers.) (2)

(Total 12)

6 Indicators

What I already know (page 44)

- Acids are liquids that may be **dangerous** if not handled carefully. (1)
- In the mouth, **bacteria** turn sugars into acids, which can cause cavities in teeth. (1)
- Acidic foods such as lemons and vinegar taste **sour**. (1)
- Strong acids can **dissolve** materials. This is called **corrosion**. (2)
- When handling acids (other than food products) you should always wear **eye protection**. (1)

Hazard symbols

(a) corrosive
(b) harmful to health
(c) harmful to the environment (3)

Indicators: wordsearch (page 45)

1. Lemon juice is a weak **acid**.
2. The opposite to an acid is an **alkali**.
3. To test the acidity of a substance we can use an **indicator** such as **litmus**.
4. A substance that is not an acid or an alkali is described as **neutral**.
5. Bicarbonate of soda dissolves in water to form an **alkaline** solution that will turn litmus **blue**.
6. Lemon juice and **vinegar** turn litmus pink.
7. Substances, such as strong acids, which 'eat into' other materials are described as **corrosive**.
8. Foods that are **acidic** taste sour. (10)

Testing common substances (page 46)

1. (a) Area A soil is neutral. (1)
   (b) Area B soil is alkaline. (1)
2. The soil from the pot is slightly acidic. (1)
3. Cabbage: A and C
   Azalea: C
   Blueberry: C
   Blackcurrants: A, B and C
   Clematis: A or B
   Mint: B (6)
4. The special indicator tells you the strength of the acidity or alkalinity, whereas litmus only gives basic information. (2)

Making an indicator (page 47)

1.
   - Step 1: Place the berries in a pestle and mortar.
     Reason: This is the apparatus used for crushing materials. (2)
   - Step 2: Add a small amount of sand to the mortar.
     Reason: To increase friction. (2)
   - Step 3: Crush the berries.
     Reason: To break down the material to extract the juice. (2)
   - Step 4: Filter the liquid.
     Reason: To remove sand and other solid material from the extract. (2)
2. Yes, because the colour changes distinguish between acidic, neutral and alkaline solutions. (2)

Quick quiz: indicators (page 48)

1. Acid/acidic, neutral, alkali/alkaline. (3)
2. Lemon juice, vinegar, other citrus fruit (Any two valid suggestions) (2)
3. Neutral (1)
4. Indicators (1)
5. Pink (1)
6. Blue (1)
7. See table below. (Allow purple for neutral solutions.) (3)

Liquid	Colour shown on pink litmus strip	Colour shown on blue litmus strip
Acid	pink	pink
Neutral substance	pink	blue
Alkali	blue	blue

8. Red cabbage (allow other valid alternatives) (1)
9. Eye protection/wash hands after handling (1)

(Total 14)

7 Light and sight

What I already know (page 50)

Allow 1 mark for each scientifically correct sentence using words or phrases from each list.

Sentences could include:

- **Light** always travels in **straight lines**.
- **Transparent** materials allow light to **pass through**.
- **Opaque** objects **block** light.
- **Mirrors reflect** light.
- A **shadow** is an area where light has been **blocked**.
- **Luminous** objects **give out** light.
- We **cannot** see things without a light **source**.
- We **see** objects when light from them **enters** our eyes.
- We should **never** look directly at the **Sun**.
- A **translucent** material does not allow you to see through **clearly**. (10)

Light and shadows (page 51)

1. They should place the shadow puppet at a distance (preferably specified) from the wall and measure the height of the shadow. They should move the puppet further from/closer to the wall in regular steps, measuring the height at each position. The lamp should remain in the same position throughout. (6)
2. Correctly drawn table. (1) The independent variable (distance from puppet to wall, in cm or m) should be in the left-hand column and the dependent variable (height of shadow, in cm) should be in the right-hand column. (1) Suitable column headings including units. (2) Results correctly entered into the table. (2)
3. Anomalous point (30, 7.2) correctly circled. (1)
4. Ideally, they should check the result by repeating the reading for that point. (Allow ignoring the result when drawing the line on the graph since the pattern in the results is clear.) (1)
5. Straight line joining points but omitting the anomalous reading. (1)
6. 7.9 cm (or value correctly read from graph as drawn). (1) Lines to show the working. (1)

Reflection (page 53)

1. Correctly drawn reflected ray at 30°. (2) Correctly labelled angle of incidence and angle of reflection. (2)
2. The treasure is in chest C. (Marks awarded at the discretion of the marker to reward accuracy and neatness.) (8)

Sight (page 54)

This is an opportunity for children to create their own appropriate success criteria. They should think about/discuss what an attractive and informative page in a magazine should look like and what would make an explanation clear and understandable for the target audience.

They might set the following criteria:

- eye-catching to attract the reader/have an eye-catching, interesting title (1)
- illustrated by a neat diagram (1)
- use examples that appeal to the reader (1)

(Allow other suitable criteria.)

The concept to be explained is: we see non-luminous objects when light from a light source is reflected off the object into our eyes.

This should ideally be accompanied by a correctly drawn diagram, e.g.:

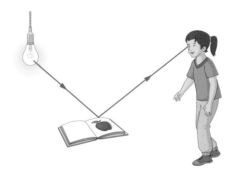

(3)

In a classroom setting this could usefully be peer assessed.

Light and sight: crossword (page 55)

Across
3 protractor
6 sunglasses
7 incident
9 opaque
10 reflects

Down
1 straight
2 periscope
4 transparent
5 luminous
8 normal
11 shadow (11)

Quick quiz: light and sight (page 56)

1 Luminous (1)
2  (1)
3 Reflection (1)
4

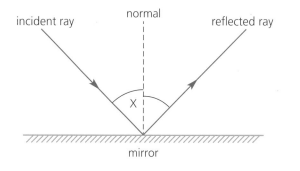

(4)

5 X marks the angle of incidence (see diagram above). (1)
6 The angle of incidence is equal to the angle of reflection. (1)
7 Periscope/some telescopes/rear view mirrors in cars/ kaleidoscope/mirrors at road junctions or driveways (Allow suitable alternatives.) (1)
8 The body is **opaque** and **blocks the light**. (2)
9 Similarity: same shape (1)
  Difference: colour, size, surface features
  (Allow suitable alternatives.) (1)
10 Neatly drawn diagram showing light ray from light source striking the object and being reflected to the observer's eye. For example: (2)

(Total 16)

8 Changing electrical circuits

What I already know (page 58)

Electrical components will only work if they are connected into a **complete** circuit. The circuit must contain one or more cells to provide the electrical **energy**. A circuit that has a single pathway for the electricity to flow around is described as a **series** circuit.

To make a lamp glow more brightly we can add more **cells**. If we add more lamps they will glow **less** brightly. A **switch** is a device that makes a break in a circuit to stop the flow of electricity.

Materials that allow electricity to pass through them are called electrical **conductors**. Nearly all **metals** are good conductors. Non-metals are mostly **insulators**. One non-metal that is a good electrical conductor is **carbon** (**graphite**).

Mains electricity can be **dangerous** if not used carefully. The cables on household items that use electricity to work, such as a **kettle**, an iron or a **lamp**, will have a layer of an insulating material such as **plastic** or rubber around them. For safety, these items will also have a **fuse** in the plug to cut off the electricity if a fault occurs. (15)

Changing circuits: graphic organiser (page 59)

Parts of the circuit:
- (Cell) Switch (Wires) Lamp (2)

What will happen if the part is missing:
- Cell – the lamp will not light because there is no energy flowing round the cell. (2)
- Switch – The lamp cannot be switched off because there is no way of breaking the circuit. (2)
- Wires – The lamp will not work because no electricity can flow. (2)
- Lamp – Nothing visible will happen when the switch is closed because there is no active component (but a short circuit will be formed). (2)

Changing circuits: odd one out (page 60)

1 B is the odd one out because the circuit does not contain a switch. (The position of the cells in circuits A and C makes no difference to the way the circuit works.) (2)
2 F is the odd one out because the bulb will not light, because it is not connected properly (one of the wires should be on the tip of the bulb). (2)
3 H is the odd one out because the cells are connected the other way round, so the motor will turn in the opposite direction to the others. (2)

Drawing circuit diagrams (page 61)

1 The circuit should show a motor, a switch and one cell connected in series. (Marks could be awarded for neat drawing.) (5)
2 The circuit should show two lamps, a switch and three or four cells. (5)
3 The circuit should show a switch, a lamp and possibly a buzzer. There should be one or two cells if only a lamp is shown and additional cells if a buzzer is included in the circuit. (5)

Short circuits (page 62)

1. The bulb will not light because the circuit contains a short circuit. This might become very hot and is therefore potentially dangerous. (3)
2. There are too many cells, and the lamp will be overloaded. It might create a flash, and there is a slight possibility that the glass envelope might shatter. (3)
3. The cell could be dead, the cell is not producing sufficient energy to light two lamps, one of the lamps might be broken or there might be a loose connection in the circuit. (4)

Changing electrical circuits: puzzle (page 63)

1. symbol
2. lamp
3. series
4. energy
5. component
6. circuit
7. motor
8. buzzer
9. conductor
10. cell
11. insulator (11)

The hidden word: electricity (1)

Quick quiz: changing electrical circuits (page 64)

1. (a) Normal brightness (1)
   (b) The lamp would be brighter. (1)
2. The lamp would be dimmer. (1)
3. Correctly drawn symbols for lamp, buzzer, cell, motor and switch (5)

switch (open)

4. A circuit where all the components are linked into a single pathway (1)
5. (a) Short circuit (1)
   (b) Short circuits can produce heat/might cause a fire. (1)

(Total 11)

Evolution

Skill Using and applying scientific knowledge
Creativity

The picture shows an undiscovered island in the middle of an ocean. The island is very marshy with lots of streams and lakes containing many fish. Some areas of dry land have a few trees and fruiting bushes growing on them, and insects scuttle around in the undergrowth. It is very hot during the daytime, but becomes very cold at night.

Imagine that Noah's Ark ran aground on a sandbank near the island. One pair of animals escaped from the ark and was left behind on the island.

Choose one type of creature that might have been on the ark and imagine how that creature might evolve over millions of years to become well adapted to the conditions on the island. Think about the life processes (movement, feeding, reproduction and growth) and imagine how the animals might carry these out on the island.

Draw a picture of your evolved animal and label it to show the changes that have occurred.

Skill check

I have used my scientific knowledge to suggest adaptations that would be needed. ☐

I have drawn and labelled a picture to show my ideas clearly. ☐

Making new varieties of plants and animals

Skill Using information to make predictions

Dog breeders consider lots of features of different breeds when deciding which to choose as parents of their puppies. In 1989, an Australian dog breeder called Wally Conron needed to train a guide dog for a person who was allergic to the hair of many dogs. He chose to cross breed a Standard Poodle and a Labrador. Some information about these two breeds is given in the table.

	Hair shedding	Affectionate	Intelligence/ trainability	Barking	Sensitivity	Aggressive behaviour
Labrador	High	Very	High	Noisy	Easy-going	Low
Poodle	Very low	Very	High	Quiet	Nervous	Low

1. One of the puppies (the first labradoodles to be born) had the right characteristics to be the guide dog. He was named Sultan. Complete the table below to suggest the value for each characteristic that Sultan might have shown.

	Hair shedding	Affectionate	Intelligence/ trainability	Barking	Sensitivity	Aggressive behaviour
Guide dog						

(6)

2. It is likely that Sultan was not the only puppy in the litter. Would you expect all the puppies to show the same characteristics? Explain your answer.

_____

_____

(2)

3. Suggest characteristics that breeders might be looking for in dogs for the following purposes:

   (a) Detecting drugs in airports

   _____ (2)

   (b) Family pet

   _____ (2)

   (c) Pulling a sled

   _____ (2)

   (d) Helping a deaf person

   _____ (2)

Skill check

I have used information to predict possible outcomes of cross breeding. ☐

Evolution: puzzle

Skill Developing scientific vocabulary

Use the clues below to help you to fill in the puzzle.

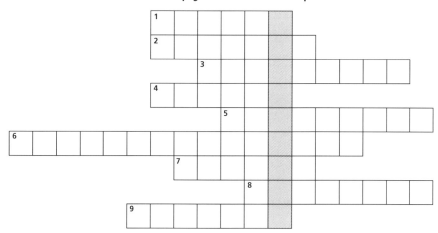

Clues

1 The fossil hunter from Lyme Regis who first discovered a fossil *Ichthyosaurus*
2 A group of living things that are able to breed successfully with each other
3 The group of reptiles that lived on Earth between about 230 million years ago and 65 million years ago
4 The preserved remains of an animal or plant that died millions of years ago
5 Changes in animals or plants over many generations resulting in the formation of a new species
6 A scientist who studies fossils
7 The scientist whose book *On the Origin of Species* first described how evolution occurs
8 One of these celestial bodies is believed to have contributed to the extinction of the dinosaurs
9 The word used to describe a species of which no living examples remain

(9)

The letters in the shaded squares spell out another word you may have met during this topic. Write a definition for this word. You could use a dictionary or the glossary from your textbook to look it up but then write it in your own words.

_____ (2)

Skill check

I know and understand the vocabulary for this topic.

Quick quiz: evolution and inheritance

Write *short* answers to the following questions to check how much you remember about evolution and inheritance. You do not need to write full sentences.

1 How do we know that living things have changed over millions of years?

_____

_____ (2)

2 What does a palaeontologist study?

_____ (1)

3 Who discovered the first fossils of *Ichthyosaurus*, *Plesiosaurus* and *Pterodactylus*?

_____ (1)

4 Which two scientists independently worked out how new species of living things occur?

_____ (2)

5 Explain briefly how tortoises living on an island where most of the vegetation is high off the ground might develop longer necks over a number of generations.

_____

_____

_____

_____ (3)

6 What name is given to the process of change over millions of years to produce new species?

_____ (1)

7 Explain briefly why a litter of kittens will not all be identical to their parents.

_____ (1)

8 What is meant by the term 'cross-breeding'?

_____

_____ (2)

(Total 13)

Check list: evolution and inheritance			
I know	☺	😐	☹
that living things have changed over time			
that fossils provide evidence about life on Earth millions of years ago			
how the work of palaeontologists led to ideas about evolution			
how adaptation can lead to evolution			
that living things produce offspring of the same kind			
that the offspring may vary and are not identical to their parents			
that cross-breeding is used to produce new varieties of living things			

5 Hot and cold

What I already know

Draw straight lines to match each picture with the correct temperature.

 Boiling water

200 °C

 Body temperature

37 °C

 Ice lolly

100 °C

 Hot oven

−15 °C

 Room temperature

0 °C

 Freezer

20 °C

 Fridge

5 °C

(7)

Measuring temperature

Skill Reading thermometer scales

We can measure temperatures using a thermometer.

Usually we measure temperatures using units called degrees Celsius (°C). Other scales are also used and sometimes thermometers show the scales in both °C and °F (degrees Fahrenheit).

Look carefully at the scales on the thermometer below.

1 Read the temperatures shown by the arrows. Remember to give the correct units.

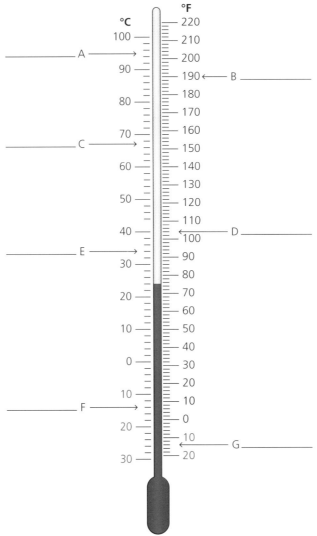

(7)

2 Using your knowledge of the properties of water and the thermometer above, find out:

(a) the boiling point of water in °F _____ (1)

(b) the freezing point of water in °F _____ (1)

3 The temperature in a domestic freezer is −15 °C. If the temperature rises by 2 °C what is the new temperature? _____ (1)

Skill check

I have read thermometer scales correctly in °C and °F. ☐

Keeping warm: potatoes

Skill Plotting a bar chart

Drawing a conclusion from experimental data

Some children are each given a baked potato from the oven. They measure the temperature of the potatoes using a temperature probe.

They have five identical shoe boxes. They select materials to put into the boxes to keep their potato as warm as possible for 30 minutes.

After 30 minutes, they measure the temperature of their potatoes again.

Material used	Temperature at start, in °C	Temperature after 30 minutes, in °C	Change in temperature, in °C
Nothing	92	49	
Woolly scarf	92	66	
Bubble wrap	92	67	
Newspaper	92	60	
Hay	92	55	

1 Calculate the change in temperature for each potato and write these in the table above. (5)

2 Suggest why they left one potato with no insulation.

_____

_____ (1)

3 Suggest one thing that they should do to make sure that their investigation is a fair test.

_____

_____ (1)

4 On the axes, draw a bar chart to show the change in temperature for each material. (5)

5 Which material is the best insulator? Explain how you can tell.

_____

_____ (2)

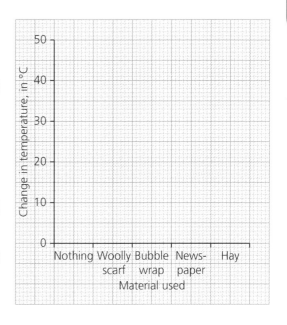

Skill check

I have drawn a neat bar chart, using a pencil and ruler to draw the bars. ☐

I have accurately plotted the relevant data. ☐

I have used the data effectively to draw a conclusion for this investigation. ☐

Keeping warm: house insulation

Skill Evaluating evidence

The table below gives some information about three materials that might be used to insulate a building.

Material	Cost per m²	Pros (+) and cons (−)
Cellulose fibre Insulation properties: Good	£4	(+) Made from recycled newspaper and card Absorbs heat in warm weather and releases it when conditions are cool
		(−) Contains printing inks which may release vapours into the air May settle over time, reducing air pockets Possible fire risk if not treated with fire-retardant chemicals
Sheep wool Insulation properties: Good	£9.50	(+) Renewable Excellent at absorbing moisture Long lasting
		(−) Sheep produce methane, which contributes to climate change
Straw Insulation properties: Moderate	About £1	(+) Renewable Provides habitat for animals whilst growing Widely available
		(−) Will decay over time, especially if damp May encourage insects and rodents Possible fire risk if not treated with fire-retardant chemicals

1 Which of the materials is the most expensive to install? _____ (1)

2 Which of the materials is the worst thermal insulator? _____ (1)

3 What evidence suggests that the cellulose fibre insulation might be a good choice in a climate with hot days and cool nights?

_____

_____ (2)

4 From the evidence in the table, which of the products would you expect to last longest after installation? Explain your answer.

_____

_____ (2)

5 All of these materials claim to be environmentally-friendly. For each material, underline one piece of evidence in the table that supports this claim and write an X by one piece of evidence that does not support the claim. (6)

Skills check

I have interpreted information from the table to answer questions. ☐

I have identified evidence to support and refute a claim. ☐

Research topic: polar bears

Skill Researching a topic using books and the internet

Keeping a bibliography

Polar bears are very well adapted to survive the extremely harsh conditions in the Arctic. Use books or the internet to help you to find out about how polar bears keep warm in these very cold conditions. Write your findings in your own words and keep a record of the book titles and websites that you use for your research. You might like to include a labelled picture or diagram to illustrate your work.

_____
_____
_____
_____
_____
_____
_____
_____

Bibliography

_____
_____
_____

Skills check

I have found reliable sources of information in books and/or the internet. ☐

I have recorded my sources of information in a bibliography. ☐

Quick quiz: hot and cold

Write *short* answers to the following questions to check how much you remember about hot and cold. You do not need to write full sentences.

1 What name do we give to a material that does not allow heat energy to pass through easily?
_____ (1)

2 Why are feathers often used in duvets, sleeping bags and winter jackets?
_____
_____ (2)

3 Explain how an insulated cool bag works.
_____ (1)

4 What is meant by the term 'temperature'?
_____ (1)

5 What piece of equipment is usually used to measure temperature?
_____ (1)

6 What units are usually used when measuring temperature?
_____ (1)

7 What is the boiling point of water?
_____ (1)

8 What is the freezing point of water?
_____ (1)

9 What is the normal body temperature of a healthy human?
_____ (1)

10 Give two ways in which we can reduce loss of heat energy from our houses.
_____
_____ (2)

(Total 12)

Check list: hot and cold

I know	☺	😐	☹
that a thermal insulator is a material that does not allow heat energy to flow through easily			
that some materials are better thermal insulators than others			
that air trapped within materials is a good insulator			
that temperature is a measure of how hot or cold things are			
that temperature is measured using a thermometer			
that the units of temperature are °C			
that water freezes at 0 °C and boils at 100 °C			
that the temperature of a healthy human body is 37 °C			

I can	☺	😐	☹
use a thermometer safely to measure temperatures			
read temperatures below 0 °C on a thermometer scale			
use a variety of sources to research a topic			

6 Indicators

What I already know

Use words from the box to complete the following sentences. Each word may be used once, more than once or not at all.

| bacteria | dangerous | eat | safe | sour |
| corrosion | dissolve | eye protection | saliva | sweet |

Acids are liquids that may be _____ if not handled carefully. (1)

In the mouth, _____ turn sugars into acids, which can cause cavities in teeth. (1)

Acidic foods such as lemons and vinegar taste _____ . (1)

Strong acids can _____ materials. This is called _____ . (2)

When handling acids (other than food products) you should always wear _____ . (1)

Hazard symbols

Containers of acids and other chemicals often show hazard warning symbols. Here are three that might be seen on an acid container. Look at them carefully and decide which symbol matches each warning. Draw straight lines to link the pairs.

(a)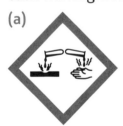

harmful to health

(b)

harmful to the environment

(c)

corrosive

(3)

Indicators: wordsearch

Skill Using and understanding scientific language

Choose words to complete the following sentences and write them in the spaces provided. Check your answers by finding the words in the wordsearch and highlighting them neatly.

1 Lemon juice is a weak _____

2 The opposite to an acid is an _____

3 To test the acidity of a substance we can use an _____ such as _____

4 A substance that is not an acid or an alkali is described as _____

5 Bicarbonate of soda dissolves in water to form an _____ solution that will turn litmus _____

6 Lemon juice and _____ turn litmus pink.

7 Substances, such as strong acids, which 'eat into' other materials are described as _____

8 Foods that are _____ taste sour

X	Q	Z	K	R	N	K	O	X	T	I	B
Z	T	A	C	P	X	B	J	K	S	W	B
P	Z	A	C	Y	S	L	Z	I	H	C	M
Y	M	L	C	I	G	U	T	W	H	O	V
V	B	K	Y	M	D	E	J	U	X	R	L
V	X	A	O	R	E	I	Y	C	N	R	Y
I	S	L	I	N	D	I	C	A	T	O	R
N	T	I	N	E	U	T	R	A	L	S	C
E	A	Z	E	A	R	V	X	V	Q	I	S
G	C	A	L	K	A	L	I	N	E	V	I
A	I	Z	Q	L	I	T	M	U	S	E	J
R	D	Q	F	H	U	Z	F	A	O	N	G

(10)

Skill check

I have selected the correct vocabulary to complete the sentences. ☐

45

Testing common substances

Skill Applying data to a real-world situation

Leonard and Katie are making a new garden. They read about how different plants like different types of soil. Some prefer acidic soil, some prefer neutral soil and some prefer alkaline soil.

Leonard uses a special test kit to find out about the soil in two areas of the garden. He puts soil from each part of the garden into a separate test tube and adds water. He then adds a few drops of special indicator solution.

1. What does this test tell them about the soil in:
   (a) area A: _____
   (b) area B: _____
   (2)

2. Katie uses the same kit to test the soil in a pot containing her favourite plant. What does this tell her about the soil in the pot?

   _____ (1)

3. Leonard makes an area (Area C) in the garden with soil suitable for Katie's plant.
   The table shows some more plants and what type of soil is best for growing them. Place ticks in the table to show the area or areas of the garden that would be suitable for each plant.

Plant	Soil type	Area A	Area B	Area C
Cabbage	Neutral–acid			
Azalea	Acidic			
Blueberry	Acidic			
Blackcurrant	Any			
Clematis	Neutral–alkaline			
Mint	Alkaline			

   (6)

4. Look at the soil test reading chart again. Suggest one reason why Leonard chose to use the special indicator solution rather than litmus.

   _____

   _____ (2)

Skill check

I have applied experimental results and data from a table to make suggestions and predictions about a real-world situation. ☐

Making an indicator

Skill Explaining steps in experimental procedure
Evaluating evidence

Many plant materials make good indicators. To make an indicator out of some berries a scientist followed the following steps. They are not in the correct order.

- Add a small amount of sand to the mortar.
- Filter the liquid.
- Place the berries in a pestle and mortar.
- Crush the berries.

1 Rewrite the steps in the correct order below and suggest why each one was carried out.

Step 1: _____

Reason for this step:

_____ (2)

Step 2: _____

Reason for this step:

_____ (2)

Step 3: _____

Reason for this step:

_____ (2)

Step 4: _____

Reason for this step:

_____ (2)

2 The liquid from the berries was added to three separate liquids.

Colour in lemon juice	Colour in pure water	Colour in bicarbonate of soda (sodium hydrogen carbonate) solution
Pinkish purple	Purple	Blueish purple

Would you consider this to be a good indicator? Explain your answer.

_____
_____ (2)

Skill check

I have given clear explanations for the steps in this procedure. ☐

I have given a clear evaluation of the results of this experiment. ☐

Quick quiz: indicators

Write *short* answers to the following questions to check how much you remember about indicators. You do not need to write full sentences.

1 What are the three groups into which solutions are classified?

   _____ (3)

2 Name two common household acids.

   _____ (2)

3 Is pure water acidic, neutral or alkaline?

   _____ (1)

4 What name is given to substances, such as litmus, that change to different colours in acidic, alkaline and neutral liquids?

   _____ (1)

5 What colour does litmus turn in acids?

   _____ (1)

6 What colour does litmus turn in bicarbonate of soda (sodium hydrogen carbonate)?

   _____ (1)

7 Complete the table below to show the results of testing liquids with litmus paper strips.

Liquid	Colour shown on pink litmus strip	Colour shown on blue litmus strip
Acid		
Neutral substance		
Alkali		

(3)

8 Name a natural material that can be used to produce an indicator.

   _____ (1)

9 Describe one safety precaution that you should take when handling acids and alkalis.

   _____ (1)

(Total 14)

Check list: indicators

I know	😊	😐	☹️
that solutions can be classified as acidic, neutral or alkaline			
that common acids include lemon juice and vinegar			
that pure water is neutral			
that bicarbonate of soda (sodium hydrogen carbonate) forms an alkaline solution			
that indicators can be used to classify solutions as acidic, neutral or alkaline			
that litmus is an indicator that turns pink in acids and blue in alkalis			
that many natural pigments can be used as indicators			

I can	😊	😐	☹️
extract the pigment from a natural material, such as red cabbage, to make an indicator			
use litmus to test liquids			
describe safety precautions to take when handling acids and alkalis			

7 Light and sight

What I already know

Use your existing knowledge and understanding to write *ten* scientifically correct sentences about light. Each sentence should contain one word or phrase from each of the two lists below, plus other words that you choose for yourself. Try to make your sentences as interesting as possible. Each word in the lists may be used once, more than once or not at all. One sentence has been done for you.

List A

light	cannot
transparent	see
opaque	never
mirror(s)	translucent
shadow(s)	sunglasses
luminous	

List B

block(s)	reflect
source	straight lines
Sun	blocked
pass through	sunlight
enters	give out
clearly	

Sunglasses protect our eyes from **sunlight**, which can damage our eyes.

_____
_____
_____
_____
_____
_____
_____
_____
_____

(10)

Light and shadows

Skill Describing an experimental method

Recording results in a table and graph

Interpreting information from a graph

Two children are investigating shadows. They shine a lamp onto the wall and use a shadow puppet to make the shadow.

1 Describe briefly how they should use the apparatus shown in the picture to investigate how changing the distance between the shadow puppet and the wall affects the size of the shadow.

_____

_____

_____

_____

_____

_____ (6)

They record their results in a notebook like this.

<u>Shadow height at different distances from the screen</u>
at 10 cm away, height = 6.4 cm
at 20 cm = 7.0 cm
at 30 cm = 7.2 cm
at 40 cm = 8.2 cm
at 50 cm = 8.8 cm

2 Draw a neat table in the space below and record their results properly. Remember to place the independent variable in the left-hand column and the dependent variable in the right-hand column.

(6)

The children show their results in the graph below. They have plotted their results accurately on the graph.

3 One of their results does not fit the pattern. Draw a ring around this point on the graph. (1)

4 What should the children do about this result?

_____

_____
(1)

5 Draw a suitable line on the graph to show the pattern in the results. (1)

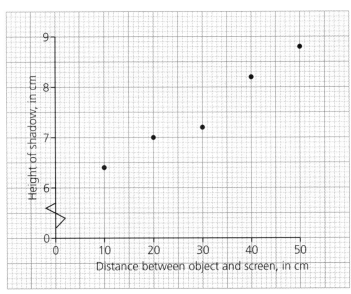

6 Use your graph to estimate the height of the shadow if the puppet was placed 35 cm from the wall. Draw lines on the graph to show your working. (2)

Skills check

I have described the experimental method clearly. ☐

I have drawn a neat table for the results with suitable column headings including the units. ☐

I have correctly interpreted data shown on a graph. ☐

Reflection

Skill Measuring angles of incidence and reflection accurately

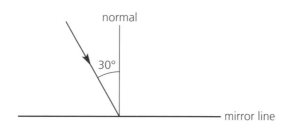

1. The diagram above shows a light ray shining onto a mirror.
   Using a sharp pencil, a protractor and a ruler, draw in the reflected ray.
   Label the angle of incidence and the angle of reflection. (4)

2. Pirate Joe is looking for his treasure. He knows that it is in one of the treasure chests. To find which is the correct chest he must use the secret mirror maze. He shines his laser pointer onto the first mirror.
   Using a sharp pencil, a protractor and a ruler, draw the path of the laser beam as it is reflected from the mirrors to see which chest contains the treasure.

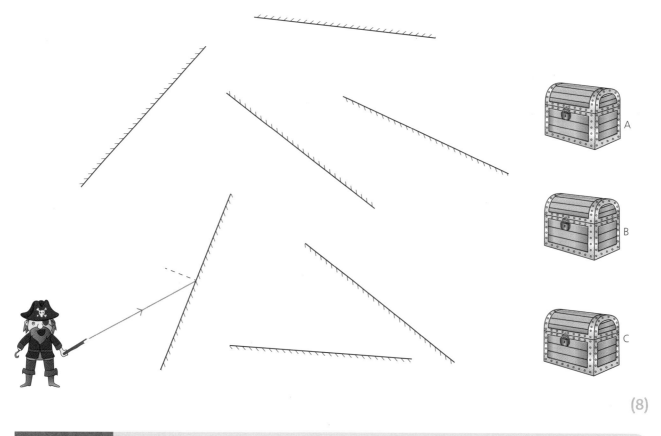

(8)

Skill check

I have used a protractor to measure angles accurately.

Sight

Skill Communicating

Setting my own success criteria

Design an article for a children's science magazine explaining how we see non-luminous objects.

(3)

Start by thinking about what might make you want to read such an article and what a clear, understandable explanation would be like. Write three things that would make your article successful.

The article must be:

1 _____ (1)

2 _____ (1)

3 _____ (1)

Skill check

I have communicated clearly how we see non-luminous objects. ☐

I have met my success criteria: 1 ☐ 2 ☐ 3 ☐

7 Light and sight

54

Light and sight: crossword

Skill Developing scientific vocabulary

Use the definitions below to help you to complete the crossword.

Across
3. The piece of equipment used to measure angles
6. A common item used to protect eyes from damage by sunlight
7. The word describing a ray of light entering a mirror
9. Blocks light completely
10. A mirror does this to light rays

Down
1. Describes the pathways along which light travels
2. A device containing two mirrors used to look over a barrier
4. Allows light to pass through unchanged
5. Producing and giving out light
8. The line constructed at 90° to a mirror surface
11. An area of darkness where light has been blocked

(11)

Skill check

I know and understand the vocabulary for this topic.

Quick quiz: light and sight

Write *short* answers to the following questions to check how much you remember about light and sight. You do not need to write full sentences.

1 What word is used to describe an object that gives out light? _____ (1)

2 Show how we represent a ray of light in diagrams.

(1)

3 What word do we use to describe what happens to a ray of light shining onto a mirror?
   _____ (1)

4 Sketch a diagram showing the path taken by a light ray falling onto a mirror. Label the incident and reflected rays. Remember to include the normal.

(4)

5 On the diagram above label the angle of incidence with an X. (1)

6 State the Law of reflection.
   _____ (1)

7 Give an example of a practical use of mirrors in a piece of equipment.
   _____ (1)

8 Explain why the human body casts a shadow in sunlight.
   _____ (2)

9 Give one similarity and one difference between a shadow and the object casting the shadow.
   _____
   _____ (2)

10 Draw a neat diagram showing how we see non-luminous objects.

(2)

(Total 16)

Check list: light and sight

I know	☺	😐	☹
that light comes from luminous sources			
that light travels in straight lines			
that light is reflected from shiny surfaces, such as mirrors			
that the angle of reflection is equal to the angle of incidence			
that a periscope is a device containing two mirrors			
that opaque objects block light to form shadows			
that the shape of a shadow is the same as that of the opaque object that casts the shadow			
that non-luminous objects can be seen when light scattered from them enters the eye			

I can	☺	😐	☹
use a protractor to measure and draw angles accurately			
explain clearly how we see non-luminous objects			
draw diagrams to show the pathway of light rays that allow us to see non-luminous objects			
describe an experimental method clearly			
draw tables to record results			
spot a result that does not fit a pattern, and suggest what to do about it			

8 Changing electrical circuits

What I already know

Use words from the box to complete the following sentences about electrical circuits.

carbon (graphite)	energy	less
cells	fuse	metals
complete	insulators	plastic
conductors	kettle	series
dangerous	lamp	switch

Electrical components will only work if they are connected into a _____ circuit. The circuit must contain one or more cells to provide the electrical _____. A circuit that has a single pathway for the electricity to flow around is described as a _____ circuit.

To make a lamp glow more brightly we can add more _____. If we add more lamps they will glow _____ brightly. A _____ is a device that makes a break in a circuit to stop the flow of electricity.

Materials that allow electricity to pass through them are called electrical _____. Nearly all _____ are good conductors. Non-metals are mostly _____. One non-metal that is a good electrical conductor is _____.

Mains electricity can be _____ if not used carefully. The cables on household items that use electricity to work, such as a _____, an iron or a _____, will have a layer of an insulating material such as _____ or rubber around them. For safety, these items will also have a _____ in the plug to cut off the electricity if a fault occurs.

(15)

Changing circuits: graphic organiser

Skill Using scientific knowledge and understanding to predict the effect of changes to a circuit

Look carefully at the picture of an electrical circuit. Think about how it works and the function of each component.

Complete the boxes beneath the picture to name the two remaining parts of the circuit and explain what will happen if each part is not in the circuit.

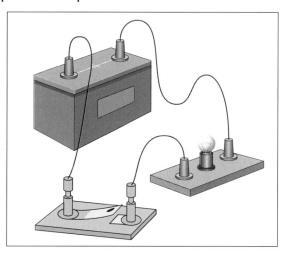

parts of the circuit

cell		wires	

(2)

What will happen if the part is not in the circuit?

(8)

Skill check

I have used my knowledge of electrical components to describe how removing them will affect the circuit. ☐

Changing circuits: odd one out

Skill Observing and interpreting circuits

In each of the following groups of circuits, one works differently from the others. Look very carefully at each circuit. Identify the odd one out in each group and explain how it is different.

1

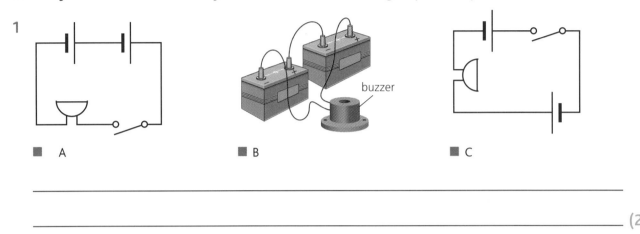

■ A　　■ B　　■ C

_____
_____ (2)

2

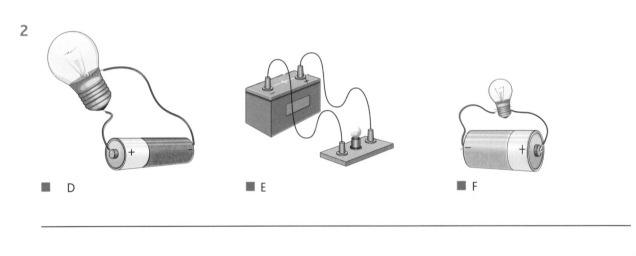

■ D　　■ E　　■ F

_____
_____ (2)

3

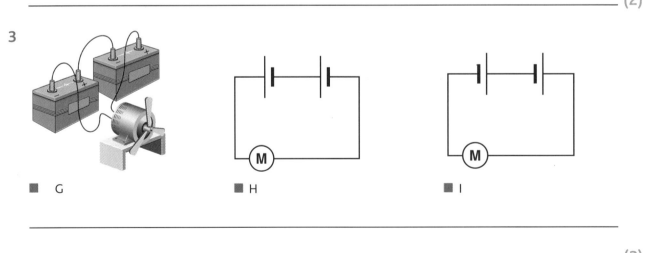

■ G　　■ H　　■ I

_____
_____ (2)

Skill check

I have identified and explained the differences between circuits. ☐

Drawing circuit diagrams

Skill Drawing neat circuit diagrams

Designing circuits to do a particular job

Electricians have to be able to design circuits to do particular jobs.

Read each of the following scenarios and draw a neat circuit diagram to show a circuit that would fulfil the need of each person.

1. Fiona has made a model windmill. She needs a circuit that will make the blades of the windmill turn slowly when she closes a switch.

(5)

2. Jamie wants to have two lamps on the front of his toy car that will light brightly when the toy driver is placed in the seat.

(5)

3. Mr Peters, who is deaf, needs a circuit that will allow him to know when someone presses the doorbell.

(5)

Skills check

I have drawn neat circuit diagrams using a sharp pencil and a ruler to draw straight lines.

I have designed circuits that would achieve the required result for each scenario.

Short circuits

Skill Spotting faults in circuits

Some children are making circuits.

Their teacher stops them because she says their circuits are not safe.

Look carefully at the diagrams of their circuits and explain why they should not be made.

1

_____

_____

_____ (3)

2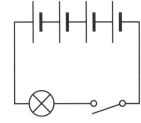

_____

_____

_____ (3)

3 Another group of children make this circuit.
They are puzzled because their lamps are not lit. Suggest what features of the circuit they should check to find out what has gone wrong.

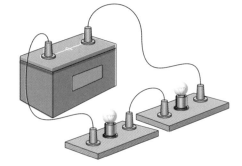

_____

_____

_____

_____ (4)

Skill check

I have identified faults in the circuits. ☐

I have explained why some faults can be dangerous. ☐

Changing electrical circuits: puzzle

Skill Knowing and understanding key vocabulary for this topic

Fill the answers to the clues into the boxes in the grid below.

Clues

1. A shape drawn to represent a component in a circuit diagram
2. A device that changes electrical energy into light
3. A word used to describe a circuit where all the parts are linked into a single pathway
4. This is provided by the cell in a circuit
5. Part of a circuit, such as a cell, lamp or switch
6. A loop containing at least one cell, wires and components around which electricity can flow
7. A electrical device that can be used to make things rotate
8. A device that makes a sound when electricity passes through it
9. A material that allows electricity to pass through it
10. The proper name for what is often called a battery
11. A material used to prevent the flow of electricity

(11)

The letters in the grey boxes in the centre can be rearranged to spell another word.

What is that word?

_____

(1)

Skill check

I know and understand the key vocabulary relating to electrical circuits.

Quick quiz: changing electrical circuits

Write *short* answers to the following questions. You do not need to write full sentences.

1 (a) How do we describe the brightness of a single lamp powered by one cell?

_____ (1)

 (b) What effect would you see if a second cell was added to this circuit?

_____ (1)

2 I make a circuit with one cell and a lamp. I then add a second lamp. What effect will this have on the brightness?

_____ (1)

3 Draw the circuit symbols for:

lamp	buzzer	cell	motor	switch

(5)

4 What is the meaning of the term 'series circuit'?

_____ (1)

5 (a) What name is given to a circuit where the electricity can flow around without doing any work because it does not flow through a component such as a lamp?

_____ (1)

 (b) Why is this type of circuit dangerous?

_____ (1)

(Total 11)

Check list: changing electrical circuits			
I know	☺	😐	☹
how changing the number of cells in a circuit can make lamps brighter or dimmer			
how changing the number of lamps or other components in a circuit can make lamps brighter or dimmer			
that a single lamp lit by a single cell glows with normal brightness			
the electrical symbols for cell, lamp, buzzer, motor and switch			
the meaning of the term 'series circuit'			
that a short circuit is one where electricity can flow without passing through a component such as a lamp or buzzer			
that a short circuit can cause heating and may be dangerous			
I can	☺	😐	☹
draw neat clear circuit diagrams using the correct symbols			
construct a circuit following a circuit diagram			
identify errors or faults in a circuit and suggest how to put them right			